The
Artful Math
Activity Book

Clarissa Grandi

About the Author

Clarissa Grandi is a mathematics teacher and geometric artist based in the UK. She is passionate about engaging learners in their mathematics through the exploration of pattern-making and mathematical art. She is the founder of ArtfulMaths.com, an online platform sharing ideas and resources to support the teaching of mathematical art and origami in schools. She can also be found on Twitter as @c0mplexnumber, where she posts and retweets creative mathematical images and ideas, and on Instagram as @clarissagrandi.art.

How to use this Activity Book

Are you a budding mathematical artist aged 9 to 99+ with a love of pattern, symmetry and color? If so, this is the perfect book for you.

This activity book is also designed to accompany the Artful Maths Teacher Book, providing teachers with a handy, ready-made resource for learners to work through in Artful Maths lessons.

Equipment

To construct the beautiful patterns in this activity book you will need the following items:

- a sharp HB pencil
- an eraser (everybody makes mistakes, even Einstein!)
- a 30cm ruler
- a pair of compasses

You may also wish to color in your patterns with coloring pencils.

Instructions

Each section of the activity book begins with an introduction to the maths behind each pattern you will be learning to draw. Then follows a series of step-by-step diagrammatic instructions. Read these carefully, ensuring you understand all the steps before you begin drawing your pattern.

You are provided with starting templates on which to construct your patterns. Sometimes the first few moves have been drawn in for you, to help you get started. Simply draw over these grey starter lines with your pencil and then continue the pattern.

You will also find **puzzlers** to puzzle over, **creative challenges** to try and further ideas to **explore**. You will find answers to these, where relevant, at the end of the book.

At the back of the activity book there is a **basic constructions** section with instructions for constructing equilateral triangles, squares and hexagons using a ruler and compass. These will be useful if you fancy the challenge of designing your own patterns from scratch in future.

At the end of the book there is a **glossary** that explains the mathematical terms used in the descriptions and instructions. The words in the glossary are **highlighted in red** the first time they appear in the activity book.

Finally, if you wish to experiment with drawing more patterns and trying out more color schemes, downloadable copies of all the templates are available on the Tarquin website (see inside back cover).

Happy mathematical art making!

Contents

1 Curves of Pursuit

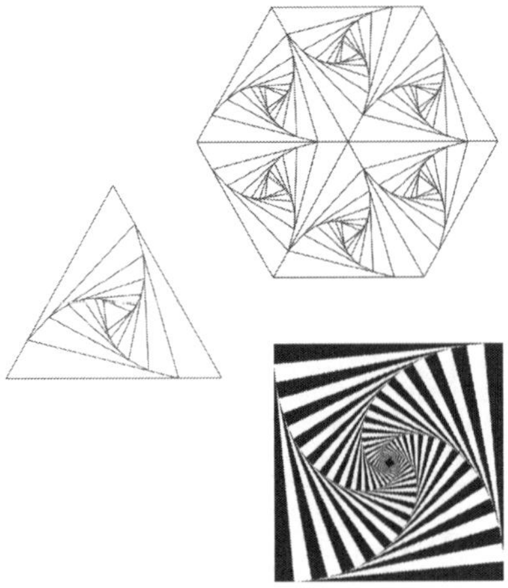

You will need:

- A pencil and a straight-edge

Optional:

- Colored pencils or pens to color in your **patterns**
- A pair of **compasses** if you wish to draw your patterns from scratch in the future (see the section on basic constructions).

What are they?

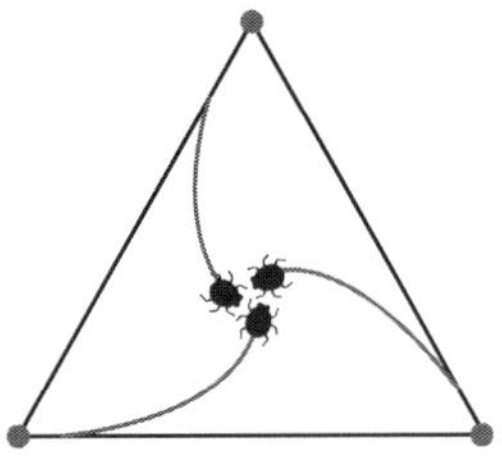

Imagine three hungry, predatory bugs sitting at the **vertices** of an **equilateral triangle**. All at once, each bug begins crawling with equal speed directly toward the bug on its right. The diagram shows the path of each bug.

The paths traced out by the hungry bugs are called **curves of pursuit**. As each bug pursues its prey, it is in turn being pursued by another bug, and as a result of this continual change in direction, lovely mathematical 'whirl' patterns emerge. The bugs can start their pursuit at the vertices of any **polygon**. Visit *http://thewessens.net/ClassroomApps/ Main/hungrybugs.html* to see them in action in a **square**.

The following rules are in place for our pursuit curves: the bugs start at the vertices of a polygon; each bug travels in the same direction (all bugs travel clockwise ↻, or all bugs travel counter-clockwise ↺); and all the bugs travel at the same speed.

Curves of pursuit are drawn by plotting the bugs' sightlines after each step they take. We will be drawing ours in a square, an equilateral triangle, a **pentagon** and a **hexagon**. We will start with the square:

STEP 1 Create your square by joining up the dots that mark the vertices of the square on the next page.

STEP 2 Next measure 1 cm to the **right** of each vertex and make a small dot. This creates the first set of footprints.

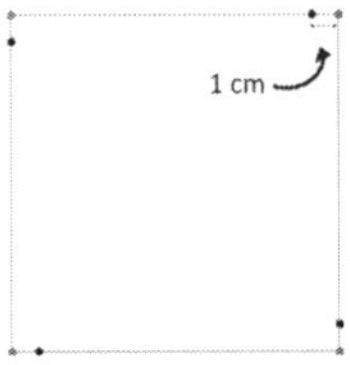

1 cm

STEP 3 Join up these four dots to make a new, smaller, tilted square that sits inside the larger original square. This is the first set of bug sightlines.

STEP 4 Now, measure 1cm to the right of each vertex **on this new square** and mark four more small dots again. This is the next set of footprints.

STEP 5 Join up this new set of dots to make an even smaller tilted square: the next set of sightlines.

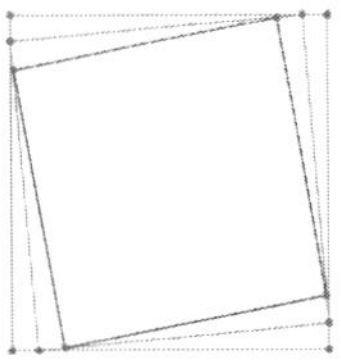

STEP 6 Repeat Steps 4 and 5 to continue the pattern until you reach the smallest square you can draw in the middle of the pattern.

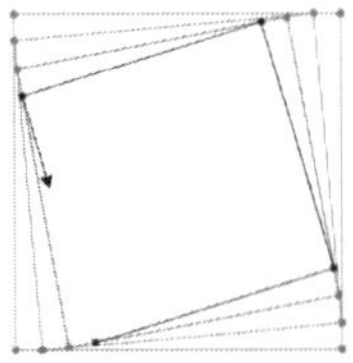

Draw your square-based curve of pursuit on the template below. The first few steps have been drawn in to get you started.

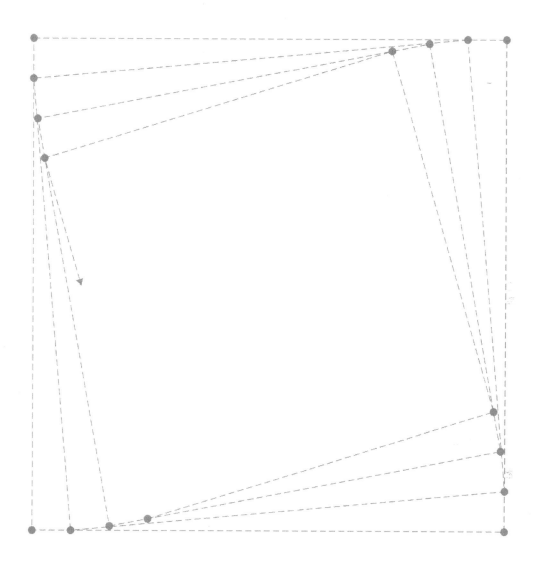

Once you have finished your square-based curve of pursuit, you can colour it in. Perhaps the colors of the rainbow, or some monochrome stripes...

Puzzler: Does the **angle** turned by the bugs after each step taken stay the same throughout the pursuit? Or get bigger? Smaller? Can you explain your answer?

Next follow the same steps to construct a curve of pursuit in the triangle, the pentagon and the hexagon templates on the following pages. Remember, if you want to experiment with different coloring schemes, there are printable copies of these templates in the downloadable resource that accompanies the activity book.

Explore: What happens if you vary the step lengths? Try 0.5cm step lengths for more detailed patterns. Experiment with using shading to create a **3D** effect.

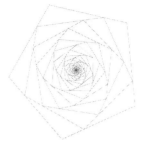

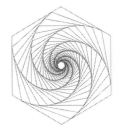

Explore: If you divide your hexagon up into 6 equilateral triangles, what happens when you alternate the direction of twist in **adjacent** triangles? Use the template on the next page to explore this.

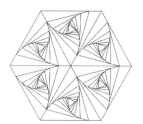

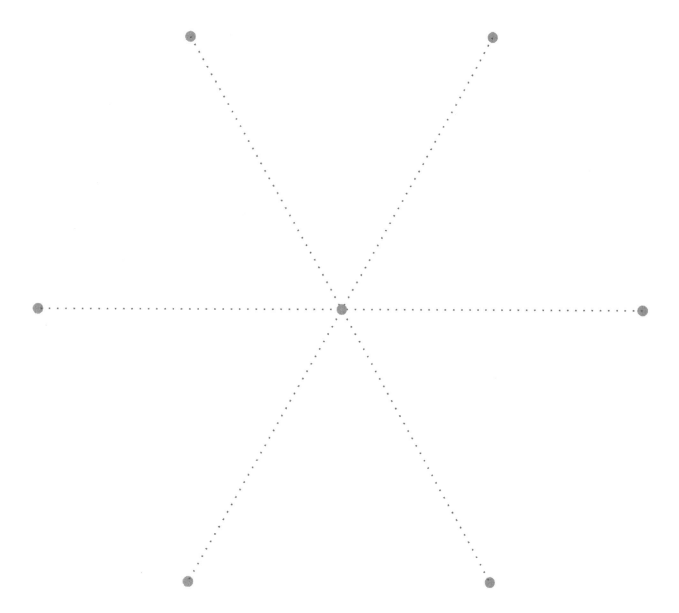

Creative challenge: Make 3D versions by printing blank **polyhedron nets** onto card (there are lots of free printable nets online) and decorating each **face** with a curve of pursuit before scoring the fold lines and assembling them. Hang them up to make beautiful mobiles.

For extra challenge, construct your own nets with straight-edge and compass. The **basic constructions** section at the back of this book will help with this.

2 Impossible Objects

You will need:

- A nice, sharp pencil, a straight-edge and an eraser
- A pair of compasses

Optional:

- Colored pencils or pens to shade or color in your patterns afterwards

What are they?

An **impossible object** is an object that can be drawn on a **2-dimensional** page to look as if it is **3-dimensional**, but which is impossible to make in reality. They are a very clever form of optical illusion.

Look at the images below: can you see why they can't really exist in 3-dimensions?

We will be learning to draw three such objects: the **Penrose triangle**, the **impossible rectangle** and the **blivet** (sometimes called the impossible trident).

How to draw the Penrose triangle

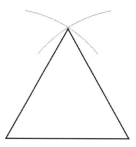

STEP 1 Using the instructions in the Basic Constructions section, construct an equilateral triangle on the base below.

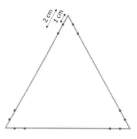

STEP 2 Measure 1 cm and 2cm from each vertex and mark these points.

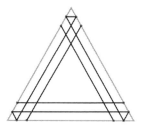

STEP 3 Lightly join these points up to make three intersecting sets of **parallel** lines with three mini triangles at each vertex.

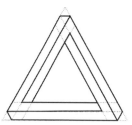

STEP 4 Using this picture as a guide, darken the lines you want to keep and erase the rest.

Your **Penrose triangle** is complete! Now use shading to make it look 3-dimensional.

    © Clarissa Grandi and Tarquin

How to draw the impossible rectangle

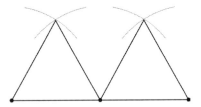

STEP 1 This time construct two identical equilateral triangles on the given 10 cm base below.

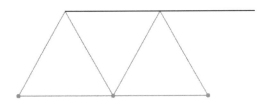

STEP 2 Draw a line across the top of the two triangles the same length as the base.

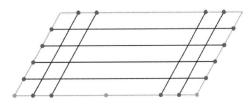

STEP 3 Join the end of this line to the base to create a **parallelogram**. Then erase the inner lines.

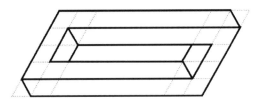

STEP 4 Now mark these points starting 1 cm from each vertex. Place the points 1 cm apart.

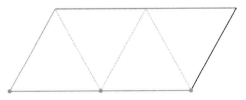

STEP 5 Lightly join the points up to create the above grid.

STEP 6 Using this picture as a guide, darken the lines you want to keep and erase the rest.

Your **impossible rectangle** is complete! Now use shading to make it look 3-dimensional.

How to draw the blivet

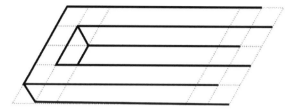

STEP 1 Start with the 'Impossible Rectangle' grid and continue up to **STEP 5**. Then darken these lines.

STEP 2 Add the rounded ends and erase the rest of the lines. Count the prongs: how many? Are you sure?

Your **blivet** is complete! What can you turn it into? Here are some ideas.

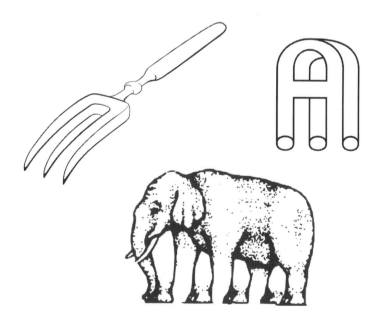

Isometric grid challenge

Can you work out how to draw the **Penrose triangle** or **impossible rectangle** on the isometric grid below? Check the answers page afterwards to see if you have got it right. Or you might like to take a quick peek right away if you need a hint to get started.

Creative challenge: Photographing an **optical illusion.** Can you arrange some wooden blocks or dice and photograph them from an angle that makes them look as if they create a 3D Penrose triangle?

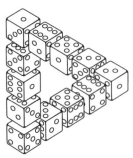

Explore: The mathematical artist M.C. Escher was famous for his drawings of impossible buildings. Look up his work: can you can see why his buildings are impossible?

3 Perfect Proportions

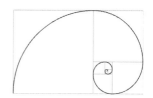

You will need:

- A pencil, a straight-edge and a pair of compasses

Optional:

- Colored pencils or pens

What is the perfect proportion?

Which **rectangle** in the group on the right do you like best? The bottom left-hand rectangle is said to be perfectly proportioned: its side lengths are in the **Golden Ratio**.

The Golden Ratio, or **Divine Proportion**, is called **Phi** (φ), and is said to occur when the **ratio** of the *smaller to the larger* of two parts is the same as the ratio of the *larger part to the whole*. In this diagram the grey part is Phi times bigger than the red part, and the whole (black) part is Phi times bigger than the grey part:

$$\left| \rule{0pt}{20pt} \right. \times \varphi = \left| \rule{0pt}{20pt} \right. \quad \text{and} \quad \left| \rule{0pt}{26pt} \right. \times \varphi = \left| \rule{0pt}{26pt} \right.$$

Phi, the Golden Ratio, is exactly $\frac{1+\sqrt{5}}{2}$, which is approximately 1.618. It is an irrational number: after the decimal point, its digits go on forever without repeating (like Pi). It can be found in lots of places in the world around us. One such place is the **Fibonacci sequence**. The Fibonacci sequence starts with 0 and 1, and continues by adding the previous two **terms** to generate the next term: 0 + 1 = 1, 1 + 1 = 2, 1 + 2 = 3, and so on. Can you continue the sequence below? Check the answers page afterwards to see if you are correct.

0, 1, 1, 2, 3, ..., ..., ..., ..., ..., ..., ...

Now, starting at 1, 1 and considering pairs of consecutive terms, divide the right-hand term by the left:

1 ÷ 1 = 1
2 ÷ 1 = 2
3 ÷ 2 = 1.5
5 ÷ 3 = ... and so on. What do you notice?

The resulting sequence of consecutive Fibonacci ratios is:

1, 2, 1.5, 1.6666..., 1.6, 1.625, 1.6153..., 1.6190..., 1.6176..., 1.6181..., 1.6179..., 1.6180... The numbers are getting closer and closer to Phi! The wonderful thing about this is that we can draw a perfectly proportioned **spiral**, the **golden spiral**, using the Fibonacci numbers.

Drawing your golden spiral

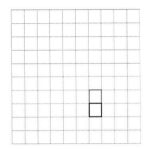

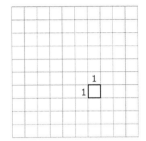

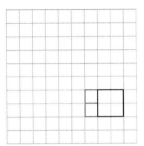

STEP 1 The first Fibonacci number after zero is 1. Begin by drawing a square of side length 1.

STEP 2 Then below it, place a square with the side length of the next Fibonacci number, which is 1 again.

STEP 3 Working counter-clockwise ↺ place the next Fibonacci square of side length 2. It fits neatly alongside as its side length is the sum of the two previous sides.

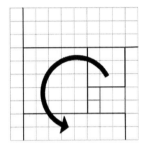

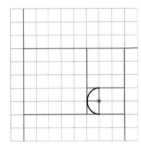

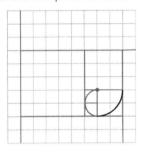

STEP 4 Continue adding squares with consecutive Fibonacci side lengths, winding your way around the growing anticlockwise spiral of squares.

STEP 5 Next, to create the spiral, use your compasses to construct quarter **circles** in each square, positioned so that the spiral is continuous.

STEP 6 Keep drawing ever larger quarter circles in consecutive squares.

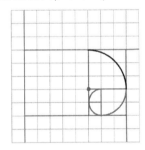

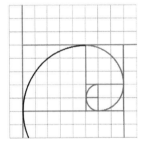

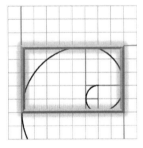

STEP 7 Think carefully about where to place your compass point each time.

STEP 8 Continue until every square is filled with a quarter circle. Your spiral is complete!

The squares in the golden spiral combine in groups to create a series of Golden Rectangles, whose side length ratios grow ever closer to Phi.

Creative challenge: Try combining several Golden Spirals to make new designs. Squared paper is available for you to download and print for this purpose. Here are some ideas for inspiration.

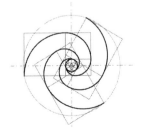

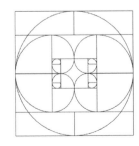

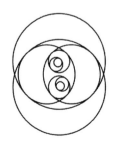

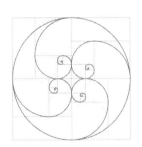

Construct your golden spiral on the grid below. The first three squares and quarter circles have been drawn in to get you started.

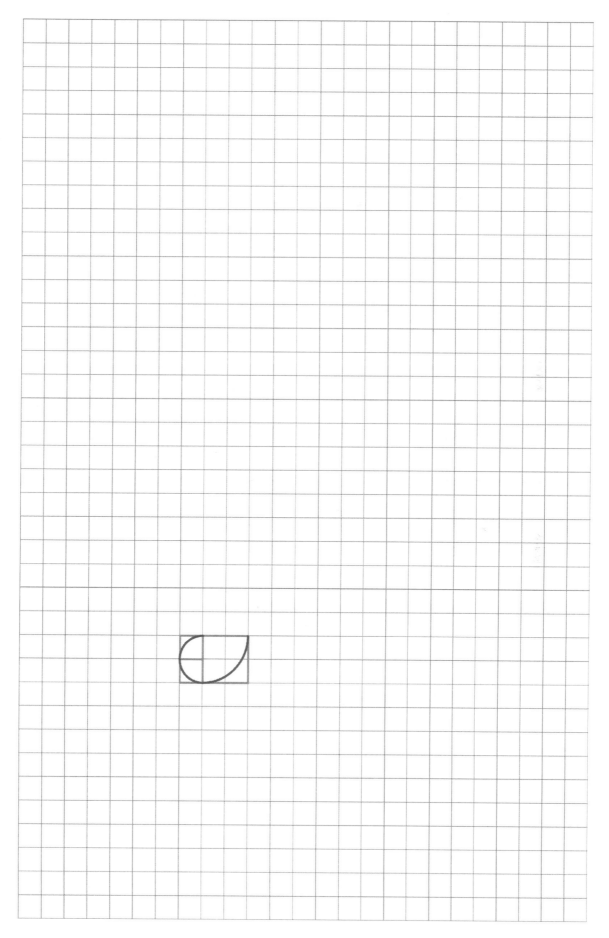

The Fibonacci sequence also appears in spiral **phyllotaxis**: the arrangements of leaves, seeds and petals in some plants. Look at the two diagrams below: you can see two sets of spirals – one set twisting counter-clockwise and the other clockwise. The number of individual spiral strands in each set are the consecutive Fibonacci numbers 8 and 13.

 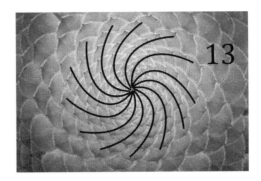

Count the spirals in the following images. Are there any hidden Fibonacci numbers?

Explore: When you are next outside, look for other examples of spiral phyllotaxis in nature. Take some photos and see if you can count the spirals.

4 Mazes and Labyrinths

You will need:

- A pencil, a straight-edge and a good eraser
- A pair of compasses for drawing the labyrinths

What are they?

Mazes are structures designed to get lost in! They have many branching paths, most of which are dead-ends, and only one true path that leads to the center or exit. **Labyrinths**, in contrast, are kinder. They are single-pathed (unicursal) and will always guide the walker to the center.

Some of the earliest mazes and labyrinths we know of were found in Egypt and in Crete, dating back over 4000 years. One of the most famous of these is the seven-circuit Cretan labyrinth, which we shall be learning to draw. In Greek mythology the Cretan king Minos owned a labyrinth in which lurked the Minotaur – a half man, half bull creature who ate anyone it came across wandering the passages.

Mazes are still very popular today, with many large ones found in the grounds of parks and grand houses. Maize mazes, cut out of fields of corn, are great fun to explore in the summer.

Drawing your branching maze

STEP 1 Begin with a rectangle with an **odd** number of squares on each side. This example is an 11 × 15 grid.

STEP 2 Then **heavily** shade in alternate squares on every other row. These squares will become the walls of your maze.

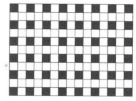

STEP 3 Now **very lightly** shade in the next set of alternate squares, but on every row this time. These squares will become either paths or walls.

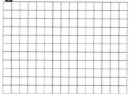

STEP 4 Make an entrance by erasing one of the lightly shaded squares on the edge. Then continue erasing lightly shaded squares to make a winding path to an exit.

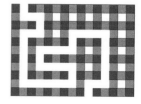

STEP 5 Now to fool people! Start creating new paths coming off this first path, but make them lead to dead ends.

STEP 6 Next heavily shade in the remaining lightly shaded wall squares. Finish by labelling the entrance and exit.

The more squares in your grid, the more complicated you can make your branching maze. Once it is complete you can test it out on friends and family.

Artful Math Handbook ISBN 9781911093176

© Clarissa Grandi and Tarquin

Draw your first branching maze on this 17 × 21 grid:

Now try drawing a more complicated branching maze on this 25 × 31 grid:

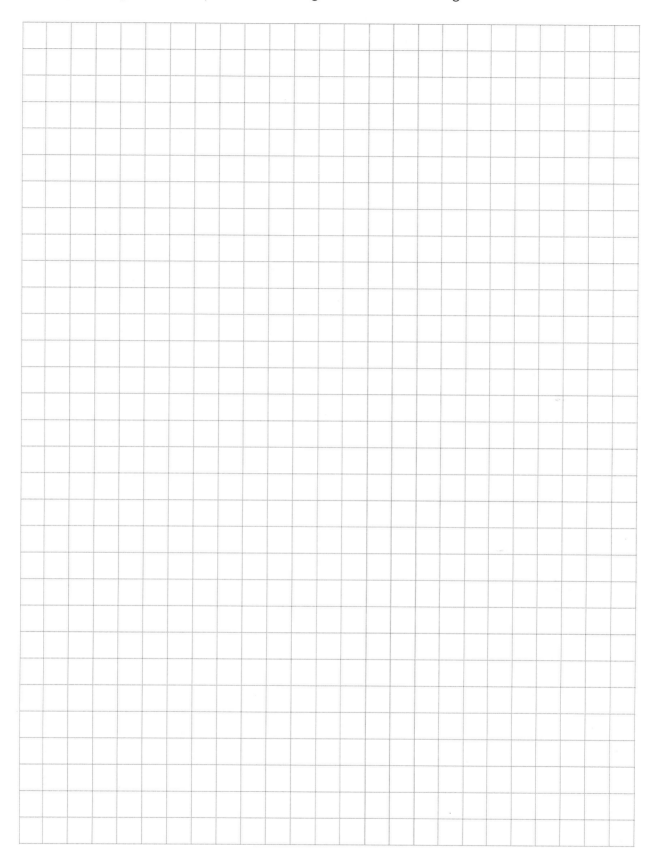

Drawing your seven-circuit labyrinth

STEP 1 Start with a set of 9 **equidistant vertical** dots (these have been drawn in for you on the next page).

STEP 2 Placing your compass point on the bottom dot, make a set of 8 concentric semi-circles.

STEP 3 Now place your compass point on the bottom *right*-hand edge of the *second smallest* semicircle and make a set of 6 concentric quarter circles that extend the original 6 outer circles.

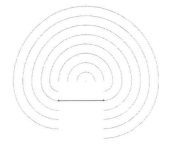

STEP 4 Move your compass point to the bottom *left*-hand edge of the *third smallest* semicircle and make a set of 5 concentric quarter circles to extend the 5 outer circles.

STEP 5 Join the edges of the second smallest quarter circles with a straight **horizontal** line.

STEP 6 Then draw a vertical straight line down from the *left*-hand edge of the smallest semicircle.

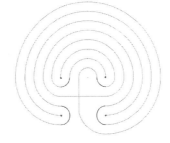

STEP 7 Place your compass point as shown and join the outermost circle to this vertical line with an **arc**.

STEP 8 Finish off by completing the remaining small quarter- and semi-circles as shown.

Your labyrinth is complete! Is there a single path to the center?

Puzzler: Can you scale your labyrinths up or down? How many concentric paths (circuits) will be in the next biggest version? Can you construct it? Can you predict the number circuits in the 5th labyrinth? The 12th? The nth?

Draw your seven-circuit labyrinth below:

.

.

.

.

.

.

.

.

.

Creative challenge: Construct an outdoor labyrinth using chalk or found objects such as stones or twigs. Or better still, draw one in the sand next time you are at the beach. You will need a length of rope or string to use as an enormous pair of compasses for drawing your arcs.

5   Epicycloids

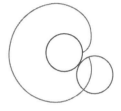

You will need:

- A pencil and a straight-edge

Optional:

- Colored pencils or pens to color in your patterns

What are they?

Epicycloids are a family of curves produced by the paths traced out by a point on the **circumference** of one **circle** when it rolls around the *outside* of another circle.

See epicycloids in action at http://mathworld.wolfram.com/Epicycloid.html

Epicycloids also appear when light is reflected off different types of curved surface. For example, when a beam of light is reflected off the inside of a circle a **cardioid** appears. This epicycloid is named after its distinctive heart-shape. You can sometimes spot one on the surface of a cup of tea or at the bottom of a glass of water!

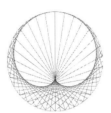 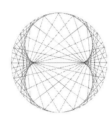 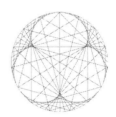

We will be using this idea of reflected rays to draw our epicycloids inside numbered circles, and we will use times tables to ensure our rays 'bounce' off the circle at the correct points. Different times tables will produce different epicycloids.

However, circles being circular, they take us back to the start again; so we will need to adjust our times tables to go back to the start too. This is called **modulo arithmetic**.

How to draw them

We start by filling out modulo times tables grids. We will be using 60-point circles, so we will start again when our numbers reach 60. Some numbers are already filled out to show the process.

Each pair of numbers in the table tell you which dots to join up with your ruler. So, for our cardioid, dot 1 joins to dot 2, dot 2 joins to dot 4, dot 3 joins to dot 6, and so on...

TIP 1: Check your working carefully before drawing your epicycloids – for example, is the final number in your table 60?

TIP 2: Cross out the numbers in the grid as you join up the pairs of dots – this will help you keep your place when you've drawn lots of lines!

To create the **cardioid**, we will map our points to the two times table, modulo 60; or '$2n$ Mod 60':

$n \to 2n$ Mod 60																			
1	2	3	4	5	6	7	8	9	10	11	12	13	14	15	16	17	18	19	20
2	4	6	8	...															

21	22	23	24	25	26	27	28	29	30	31	32	33	34	35	36	37	38	39	40
								58	60	2	4	6	...						

41	42	43	44	45	46	47	48	49	50	51	52	53	54	55	56	57	58	59	60
																		58	60

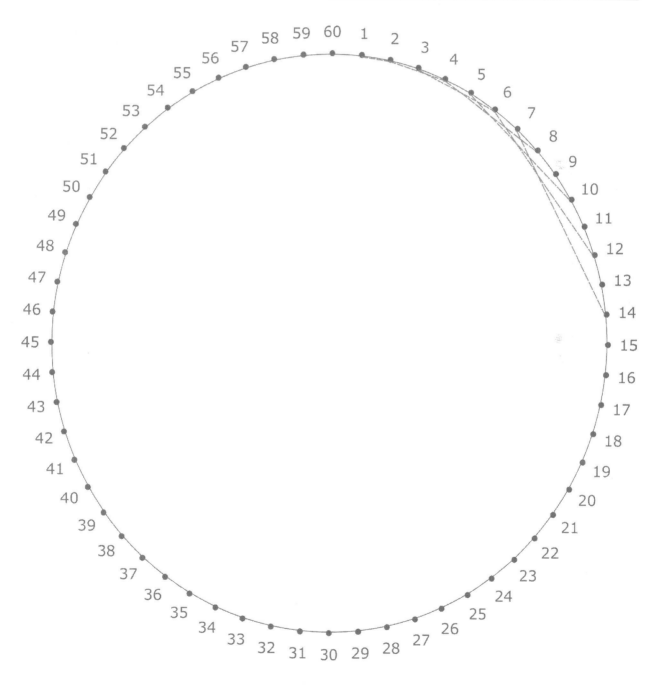

To create the **nephroid**, we will map our points to the three times table, modulo 60; or '3*n* Mod 60':

n → 3*n* Mod 60																			
1	2	3	4	5	6	7	8	9	10	11	12	13	14	15	16	17	18	19	20
3	6	9	...															57	60
21	22	23	24	25	26	27	28	29	30	31	32	33	34	35	36	37	38	39	40
3	6	...																57	60
41	42	43	44	45	46	47	48	49	50	51	52	53	54	55	56	57	58	59	60
3	...																	57	60

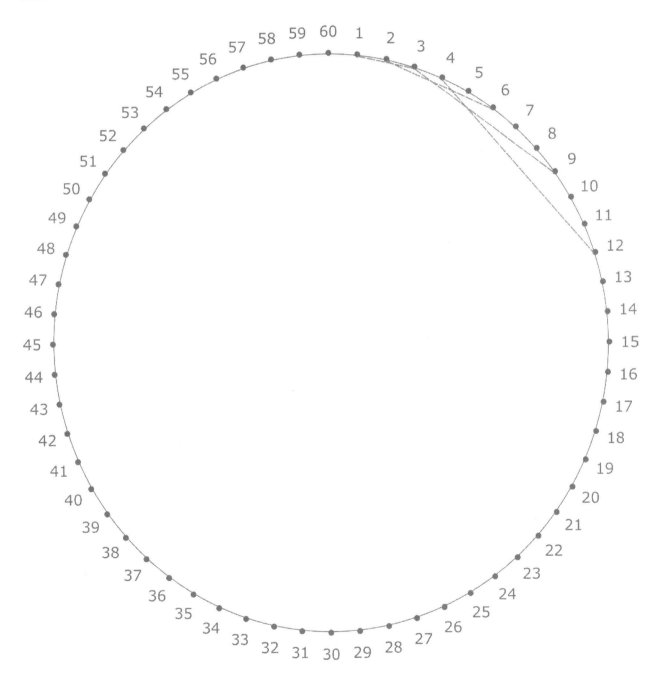

To create the next epicycloid, the **Epicycloid of Cremona**, we will map to '4*n* Mod 60':

n → 4n Mod 60																			
1	2	3	4	5	6	7	8	9	10	11	12	13	14	15	16	17	18	19	20
4	8	12	...										56	60	4	8	...		

21	22	23	24	25	26	27	28	29	30	31	32	33	34	35	36	37	38	39	40
								56	60	4	8	...							

41	42	43	44	45	46	47	48	49	50	51	52	53	54	55	56	57	58	59	60
			56	60	4	...												56	60

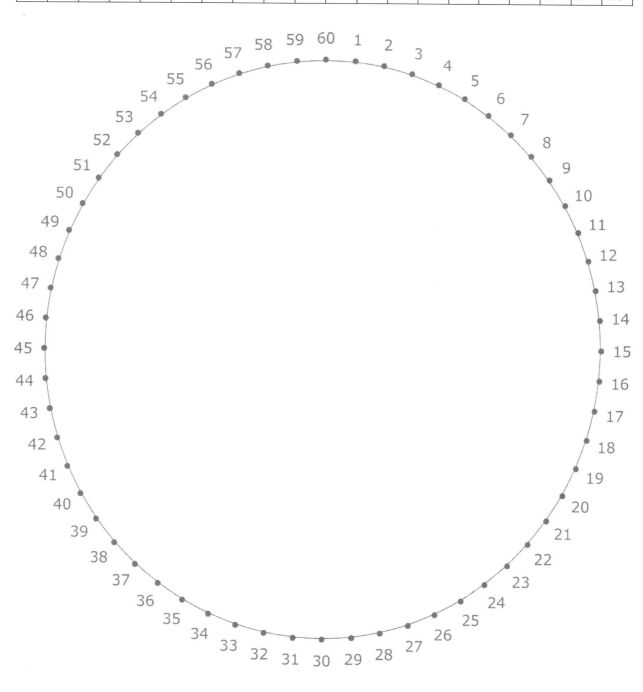

Explore: Use a flashlight to generate a cardioid inside a glass of water. What happens if you increase the number of light sources?

6 Parabolic Curves

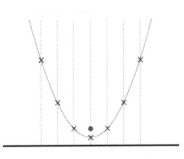

You will need:

- A pencil and a straight edge

Optional:

- Colored pencils or pens to color in your patterns

What are they?

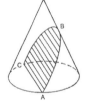

Parabolic curves can be described as a set of points that obey the rule that they are always equidistant from a straight line and a point not on the line (the **focus**). In the diagram each of the crosses is the same distance from the bold line as from the large red dot. As you can see, they trace out an elegant curve called a **parabola**.

Parabolic curves appear in other situations too. When you throw a ball, its path through the air is an approximate parabola (if it's not too windy); and if you slice through a cone at a certain angle, the cross-section is a parabola.

Parabolas have many uses. The parabola's unique shape reflects any incoming light, sound or radio waves onto its focus, which makes it a very useful shape for satellite dishes, telescopes and microphones. The parabola's shape is also perfect for supporting heavy loads over a wide span, so it is often used for arches in bridges and large buildings like cathedrals.

How to draw them

We can approximate the parabolic curve by drawing a series of straight lines between points on an **axis**.

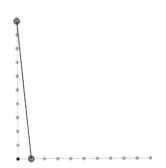

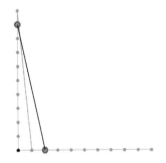

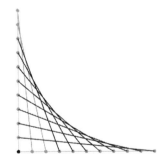

STEP 1 locate the outermost dot on one axis, and the innermost dot on the other (ignoring the center dot). Join these together with a straight, ruled line.

STEP 2 Now move one dot in on one axis, and one dot out on the other. Join these together in the same way.

STEP 3 Continue until all the points are joined and you have created your parabola.

On the next few pages you will find a set of dotted axes of increasing complexity. Explore the different designs you can make. Remember that there are more copies of these axes for you to download and print.

The basic two-way axes on which to draw your first parabola.

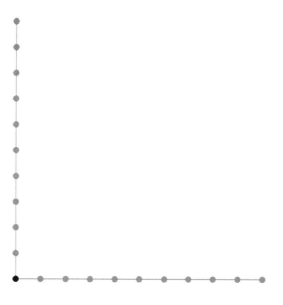

Here the basic axes have been extended to create a square. There is space for two **reflected** parabolas, or four **rotated**, overlapping parabolas.

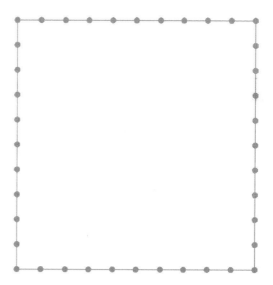

Try three overlapping parabolas rotated around this triangle.

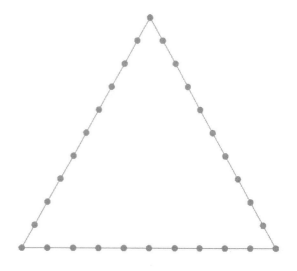

Two more sets of axes based on squares. What designs can you come up with?

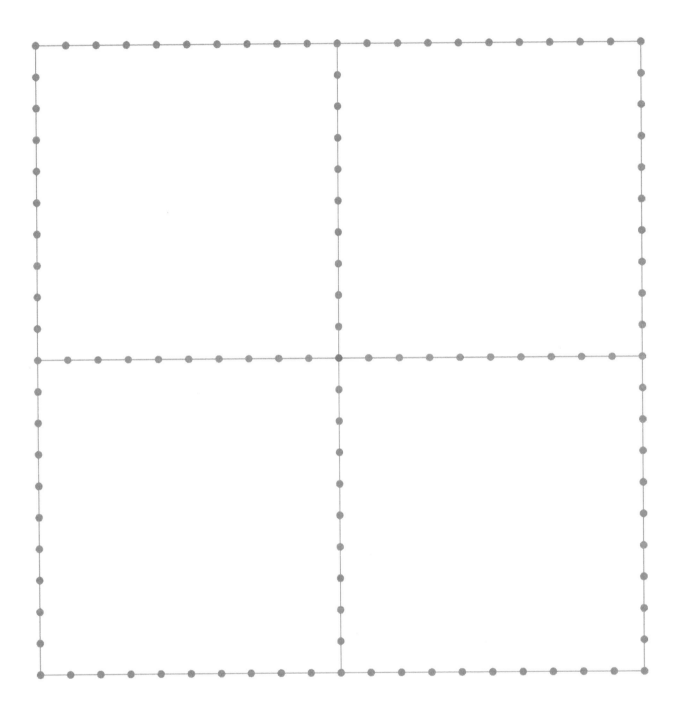

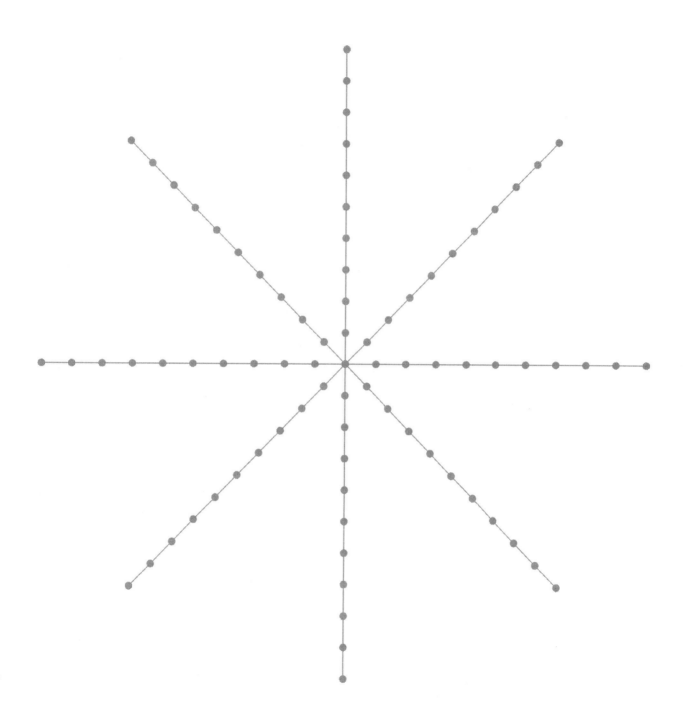

And finally, two different hexagonal axes for you to experiment with:

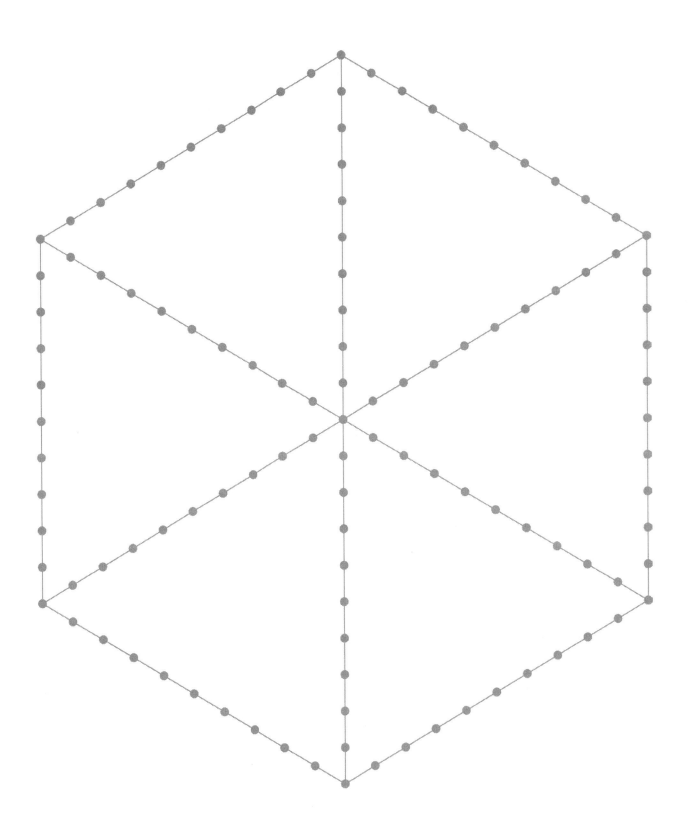

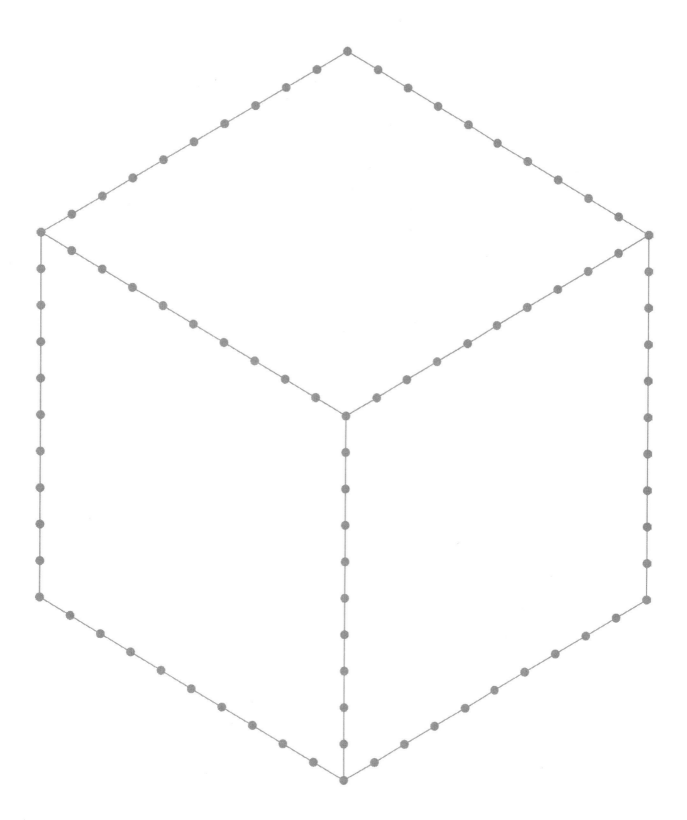

Explore: The geometric artist **Andy Gilmore** uses squared paper to draw his parabolic art. Research his work for inspiration for your own creations. If you wish to have a go, there is squared paper available for you to download and print.

The Basic Constructions

All the templates you need for your pattern-making are included in this activity book. However, once you are more confident you may wish to construct your own starting polygons from scratch.

This section shows you how to construct the three **regular** polygons that will be most useful to you: the **equilateral triangle**, the **regular hexagon** and the **square**.

Sharpen those pencils and open those compasses...

The Equilateral Triangle

Properties: All three sides are the same length and all three angles are the same size.

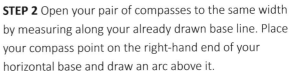

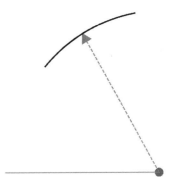

STEP 1 Decide on the side length of your equilateral triangle. Draw a horizontal line of the same width about a third of the way up your page. This will be the base of your equilateral triangle.

STEP 2 Open your pair of compasses to the same width by measuring along your already drawn base line. Place your compass point on the right-hand end of your horizontal base and draw an arc above it.

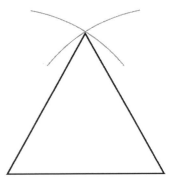

STEP 3 Without changing the opening of your compass, transfer the point to the left-hand end of your horizontal base and draw a second arc above it. The two arcs should cross each other at a single point.

STEP 4 This point is equidistant from each end of your horizontal base line. Join up each vertex to complete your equilateral triangle.

The Regular Hexagon

Properties: All six sides are the same length and all six angles are the same size.

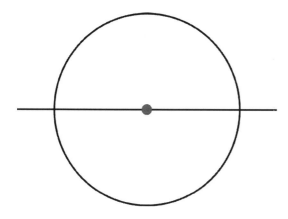

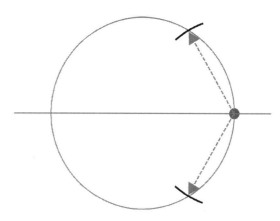

STEP 1 Draw a horizontal line about half way up your page. Place your compass point at the approximate center of this line and draw a circle. The **radius** of this circle gives the side length of your hexagon.

STEP 2 Without changing the opening of your compass, transfer the point to the right-hand intersection of your circle and the horizontal line, and make two small arcs that cross the circle above and below the line.

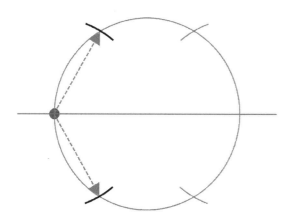

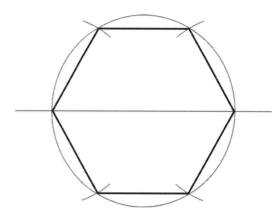

STEP 3 Now, again without changing the opening of your compass, do the same on the other side of the circle. You now have the positions of your six vertices.

STEP 4 Join the six vertices up to complete your regular hexagon. Remember, this can be subdivided into other useful shapes for your pattern making.

Regular tessellations

The equilateral triangle, the regular hexagon and the square are useful shapes for mathematical art starting points because they **tessellate**. This allows you to create repeated patterns.

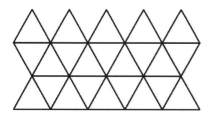

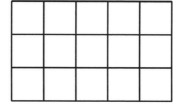

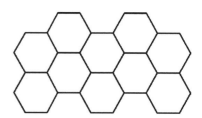

The Square

Properties: All four sides are the same length and all four angles are the same size.

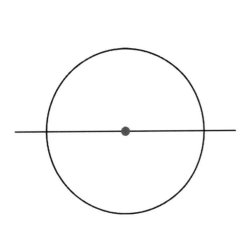

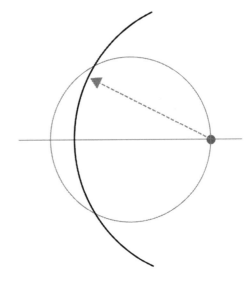

STEP 1 Draw a horizontal line about half way up your page. Place your compass point at the approximate center of this line and draw a circle.

STEP 2 Open your compasses a few centimeters *wider* than the radius of the circle. Place your compass point on the right-hand intersection and draw an arc from top to bottom.

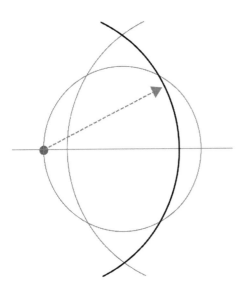

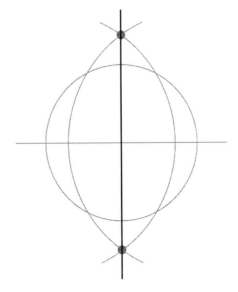

STEP 3 Without changing the opening of your compass, do the same from the other side. Check that the two arcs intersect at the top and bottom.

STEP 4 Draw a line through the two points where the arcs intersect. This is the **perpendicular bisector** of the original **diameter** of the circle.

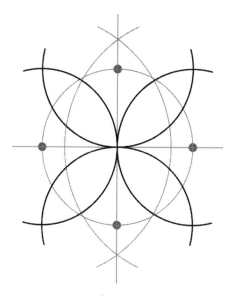

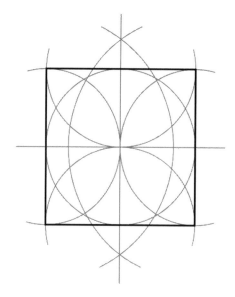

STEP 5 Now adjust your compasses back to the *original* radius of your starting circle. Place your compass point at each marked intersection and draw four more arcs, each one larger than a semi-circle.

STEP 6 The tips of the four-petalled flower shape give the positions of the four vertices. Now simply join up the vertices to complete your square.

Semi-regular tessellations

If you draw equilateral triangles, regular hexagons and squares with the same side length, then they are also able to tessellate with *each other* in repeating patterns. These are called **semi-regular tessellations**. Here are two such arrangements:

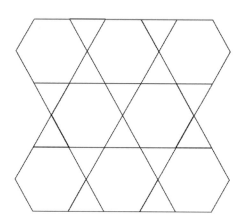

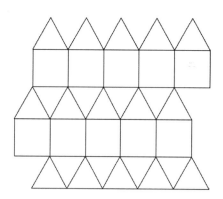

Creative challenge: There are three more semi-regular tessellations involving equilateral triangles, regular hexagons and/or squares, including one that uses all three shapes. Can you find them? Check the answers section to see if you are correct.

To help with your challenge, a printable sheet of equilateral triangles, regular hexagons and squares is provided for you to cut out and arrange. This is downloadable with the other free templates on the Tarquin Select website (see inside back cover).

Glossary

2D, 2-DIMENSIONAL: flat; having width and height but no thickness.

3D, 3-DIMENSIONAL: having width, height *and* thickness (depth); a solid object.

ADJACENT: next to each other.

ANGLE: the amount of turn between two intersecting lines, usually measured in degrees.

ARC: part of the **circumference** of a **circle**.

AXIS (plural AXES): fixed lines used for the placement of points (or coordinates).

CIRCLE: an enclosed curve that is always the same distance from the center.

CIRCUMFERENCE: the perimeter or outside edge of a **circle**.

COMPASS: an instrument with two arms used for drawing accurate **circles** and **arcs**.

DIAMETER: a straight line passing from side to side through the center of a **polygon** or **circle**.

EQUIDISTANT: the same distance apart.

EQUILATERAL: a **polygon** with all sides the same length.

FACE: a flat side on a **polyhedron**.

HEXAGON: a six-sided **polygon**.

HORIZONTAL: a line that goes from side to side across the page, **parallel** to the horizon.

NET: a flat **2D** shape that can be folded to form a solid **3D** object.

PARALLEL: two lines that are always the same distance apart and never meet.

PARALLELOGRAM: a four-sided **polygon** with opposite sides parallel and equal length.

PATTERN: images or sets of numbers that repeat, or are arranged according to a rule.

PENTAGON: a five-sided **polygon**.

PERPENDICULAR BISECTOR: a line that cuts another line exactly in two at a 90-degree right angle.

POLYGON: an enclosed **2D** shape with straight sides.

POLYHEDRON: a **3D** shape with straight edges and flat **faces**.

RADIUS: in a **circle**, the length of a straight line drawn from center to **circumference**.

RECTANGLE: a four-sided **polygon** with four **right angles**.

REGULAR: describes a polygon with all **angles** the same size and all sides the same length.

RATIO: a relationship between two quantities that shows how many times bigger one is than the other.

REFLECTION: a transformation that creates a mirror image.

ROTATION: a transformation that turns a shape around a fixed point.

SEQUENCE: a set of numbers listed in a **pattern** that follows a rule.

SPIRAL: a curve that **rotates** around its starting point getting further and further away as it grows.

SQUARE: a regular four-sided **polygon**, with four equal sides and four equal **angles**.

TERM: an individual number, or object, in a **sequence**.

TESSELLATE: fit together on a flat surface with no gaps or overlaps.

TRIANGLE: a three-sided **polygon**.

VERTEX (plural VERTICES): the corner of a **polygon** where two sides meet.

VERTICAL: a straight line that goes up and down the page, at right angles to the **horizontal**.

Answers

CURVES OF PURSUIT Puzzler (page 5) The angle turned by the bugs after each step *increases* each time. This is because the step size is always the same, but the distance between the bugs shortens each time, meaning the angle of turn must increase.

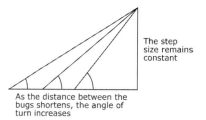

The step size remains constant

As the distance between the bugs shortens, the angle of turn increases

IMPOSSIBLE OBJECTS Isometric grid challenge (page 14):

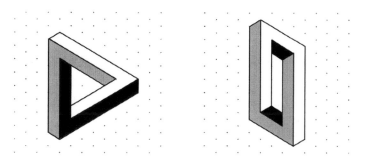

PERFECT PROPORTIONS Fibonacci sequence (page 15):

0, 1, 1, 2, 3, 5, 8, 13, 21, 34, 55, 89 ...

MAZES AND LABYRINTHS Puzzler (page 22):

1st labyrinth	3 circuits
2nd labyrinth	7 circuits
3rd labyrinth	11 circuits
4th labyrinth	15 circuits
5th labyrinth	19 circuits
12th labyrinth	47 circuits
nth labyrinth	$4n-1$ circuits

BASIC CONSTRUCTIONS Creative challenge (page 37) The remaining three semi-regular tessellations involving equilateral triangles, regular hexagons and/or squares are:

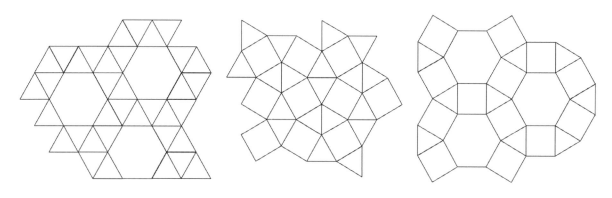

© Clarissa Grandi and Tarquin

Acknowledgements

I am grateful to Ken Wessen (@Mr_Wessen) for permission to include the link to his Hungry Bugs app at www.thewessens.net and to Kim Pitchford (@Ms_Kmp) for allowing me to use her curve of pursuit images.

My thanks to Rachel Smith and Rebekah Mellor-Read for their valuable feedback; to Judith Grandi for editorial assistance and encouragement; and to my son Joe for his sound advice.

Image Credits

Pages 3,4,6,9:   © elfinadesign/Shutterstock.com
Page 3:   © upabove/Shutterstock.com
Page 7:   © Ailisa/Shutterstock.com
Page 8:   © Ruslan Hasanov/Shutterstock.com
Page 10:   © Physicx/Shutterstock.com
Page 13:   © Igor Maltsev/Shutterstock.com
Page 13:   © Atoly/Shutterstock.com
Page 14:   © vtaurus/Shutterstock.com
Page 16:   © Anna Zasimova/Shutterstock.com
Page 16:   © Icon Font Illustration/Shutterstock.com
Page 16:   © Anna Zasimova/Shutterstock.com
Page 16:   © Anson_shutterstock/Shutterstock.com
Page 18:   © Isroi/Shutterstock.com
Page 18:   © Scot Nelson [Public domain]
Page 18:   © paperclip/Shutterstock.com
Page 18:   © aargente/Shutterstock.com
Page 18:   © photos_by_ginny/Pexel.com
Page 19:   © Roman Sotola/Shutterstock.com
Page 19:   © AnonMoos [Public domain]
Page 24:   © Wojciech Swiderski CC BY-SA 3.0
Page 28:   © Pearson Scott Foresman [Public domain]
Page 28:   © OpenClipart-Vectors/Pixabay.com

Tarquin provides many additional resources on mathematics and art – for individuals and for the classroom. You can see some of them opposite, and on our website **www.tarquingroup.com** where you can download catalogs for: Mathematics (books, posters and manipulatives), Books (including both papercraft and mathematics) and Posters. Our books are distributed in North America by IPG - search for Tarquin on www.ipgbook.com